Education 251

先有鸡还是先有蛋？

Chicken or Egg?

Gunter Pauli

[比] 冈特·鲍利 著

[哥伦] 凯瑟琳娜·巴赫 绘

何家振 译

上海远东出版社

丛书编委会

主　任：贾　峰

副主任：何家振　闫世东　郑立明

委　员：李原原　祝真旭　牛玲娟　梁雅丽　任泽林

　　　　王　岢　陈　卫　郑循如　吴建民　彭　勇

　　　　王梦雨　戴　虹　靳增江　孟　蝶　崔晓晓

特别感谢以下热心人士对童书工作的支持：

匡志强　方　芳　宋小华　解　东　厉　云　李　婧

刘　丹　熊彩虹　罗淑怡　旷　婉　杨　荣　刘学振

何圣霖　王必斗　潘林平　熊志强　廖清州　谭燕宁

王　征　白　纯　张林霞　寿颖慧　罗　佳　傅　俊

胡海朋　白永喆　韦小宏　李　杰　欧　亮

目录

Contents

在一群红原鸡中，一只雄鸡看着他的鸡群。这时，一只狐狸走了过来，对他说：

"你们红原鸡就是今天给全世界提供肉食的所有鸡的祖先？"

"是的，我听说我们是最早被驯养的禽类。如今，吃鸡肉的人比吃其他任何肉类的人都多。"

The magnificent rooster of a flock of red jungle fowl is looking at his flock, when a fox comes by and strikes up a conversation.

"So, you red jungle fowl are the ancestors of all the chickens that feed the world today?"

"Yes, I have been told that we were the first birds to be domesticated. Today more people eat chicken than any other type of meat."

一只雄鸡看着他的鸡群

The magnificent rooster looks at his flock

先有鸡还是先有蛋？

you or the egg?

"我一直想知道，先有鸡还是先有蛋？"

"这个问题已经被讨论了几千年了，人们仍然在争论！"

"那么，你个人的看法是什么？你认为是鸟还是鸟蛋先来到这个世界上？"

"I have always wondered what came first: you or the egg?"
"That question has been asked for thousands of years, and people are still debating the issue!"
"So, what is your personal take on it? Do you think it is the bird or the egg that first entered the world？"

"听着，我们从来没有想过我们的家人会在笼子里被大量饲养，更别说被迫吃玉米和大豆了。人们甚至把鱼混在我们的饲料里！"

"但是，还是回到问题的原点吧。你们以前都住在森林里，那里有很多植物和昆虫。我还是想知道，是先有鸡还是先有蛋？"

"好吧，当然先有蛋后有鸡，因为我是从蛋中出生的。"

"Look here, we never could have imagined that our family would be farmed massively in cages, let alone be forced to eat corn or soy. And with people even blending fish into our feed!"

"But let's get back to basics. You all used to live in the forest, with lots of plants and insects around. I still want to know what came first: you or the egg?"

"Well, of course the egg came before me – as I emerged from an egg."

以前住在森林里

Used to live in the forest

一定得有一只鸡下蛋

Must have been a chicken to lay it

"我再问你一次，先有鸡还是先有蛋？"

"嗯，现在我想应该先有鸡。蛋只能是鸡下的。所以，在蛋出现之前，一定得有一只鸡下蛋。"

"Let me ask you again: was the chicken first, or was the egg first？"

"Hmmm, now I think the chicken must have come first. The egg could only have been laid by the chicken. So, before there could be an egg, there must have been a chicken to lay it."

"越来越有趣了，但我还不满意。最后一次机会，你说先有鸡还是先有蛋？"

"你让我脑壳疼。这不公平！为什么不去问问你认识的其他动物呢？问问你的掠食动物朋友，比如猫头鹰或郊狼。"

"This is getting interesting. But I am not done yet. Last chance. What do you say came first: the chicken, or an egg containing a chicken?"

"You're making me think too hard. It's not fair! Why don't you go and ask some other animals you know. Ask one of your predator friends, like the owl or coyote."

比如猫头鹰或郊狼

Like the owl or coyote

你让我压力巨大

You are stressing me out

"我问你是因为我想追溯你的起源。我再问一遍，先有鸡还是先有蛋？"

"够了！你让我压力巨大。"

"拜托，放松点，用简单的逻辑想一想。还是说，你真的像我的那些狐狸朋友们认为的那样愚蠢吗？"

"I am asking you because I want to get back to your origins. So let me ask one more time. What came first: a chicken or an egg containing a chicken?"

"Enough now! You are stressing me out."

"Please, I'd like you to relax and think, using simple logic. Or are you really as stupid as many of my fox friends think you are?"

"有些人可能认为我们愚蠢，这点我很清楚。但是我告诉你，狐狸女士，至少，我们鸡是有思维逻辑的，还能表现出同理心。我们甚至还有自制力！"

　　"但是，在你告诉我先有鸡还是先有蛋之前，我是不会让你走的。"

"Some may consider us stupid, I am well aware of that. But let me tell you something, Ms Fox! At least, we chickens have logic, and can show empathy. We even have self-control."
"Still, I am not going to let you go before you tell me what came first: the chicken or the egg containing a chicken?"

我们鸡是有思维逻辑的，还能表现出同理心

We chickens have logic, and show empathy

那个蛋难道不是鸡下的吗？

Did that egg not come from a chicken?

"大家都知道，小鸡从蛋里出生，有第一只鸡之前必须先有蛋。"

"那个里面有小鸡的蛋，难道不是鸡下的吗？"

"It should be obvious to anyone that an egg, containing a chicken, had to come before the first chicken."

"And did that egg, the shell with the chick inside it, not come from a chicken?"

"好吧，可以这样推断，先有蛋，但这颗蛋是由一种不是鸡的其他生物下的！"

"对啦！谢谢你，公鸡。这是我第一次从鸡身上学到东西。我很钦佩，真的很钦佩！"

……这仅仅是开始！……

"Okay then, let's conclude that the egg came first, but it was laid by a something that was not a chicken!"

"Yes! Thank you, Rooster. This is the first time I have ever learnt anything from a chicken. I am impressed, very impressed indeed!"

... AND IT HAS ONLY JUST BEGUN!...

……这仅仅是开始！……

... AND IT HAS ONLY JUST BEGUN! ...

"先有鸡还是先有蛋"的问题最初是由希腊散文家卢修斯·梅斯特里乌斯·普鲁塔克在公元1世纪提出的。他曾写过亚历山大大帝和凯撒的传记。

The "chicken or egg question" was first posed in the first century, by the Greek essayist Plutarch (Lucius Mestrius Plutarchus). He wrote the biographies of Alexander the Great and Julius Caesar.

In the 16th century, many people thought that the first chicken did not come from an egg. This statement was fiercely questioned by the enlightened philosophers.

在16世纪，许多人认为第一只鸡不是从蛋里出来的。这种说法遭到启蒙哲学家们的强烈质疑。

In the 4th century, Aristotle wrote about the First Cause, and the debate about the origin of a chicken. He concluded that an infinite sequence exists, but with no true origin.

在公元 4 世纪，亚里士多德就"第一因"以及关于鸡的起源的争论作了论述。他的结论是，存在一个无限数列，但没有真正的起源。

The 'non-chicken' egg was produced by a 'non-chicken'. Then came the chicken, which had to be a mutation to produce the standard chicken. This involved the development of the hard shell, to protect the delicate yolk.

"非鸡蛋"是由"非鸡"下的，然后才有了鸡。必须经过基因突变才能诞生标准的鸡。这涉及硬壳的进化，壳的作用是保护脆弱的蛋黄。

Archaeological evidence suggests that the red jungle fowl was first domesticated 10,000 years ago. DNA analysis suggests that chickens diverged from jungle fowl an estimated 58,000 years ago.

考古证据表明，红原鸡最早在1万年前已被驯养。DNA分析表明，家鸡在大约5.8万年前就从原鸡中分化出来了。

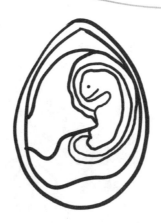

Amniotic eggs showed up roughly 340 million years ago. The egg enabled amniotes to be reproduced inside a "wet" capsule on dry land, without the need to return to water for reproduction.

大约在3.4亿年前就出现了羊膜卵。羊膜卵使羊膜动物可以在干燥陆地上的"湿"囊中繁殖，而不需要回到水中繁殖。

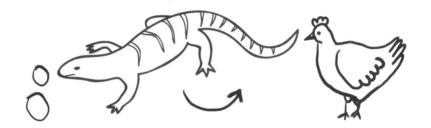

An amphibian preceded the chicken. Its egg had a gelatinous coating, and a shell membrane covering the embryo. The egg increased in size and rate of gas exchange, allowing for the embryo to develop.

两栖动物比鸡出现得早。它的卵有一层凝胶状膜，还有一层卵壳膜覆盖在胚胎上。卵被产下后，卵由小变大，气体交换频率加快，使胚胎发育。

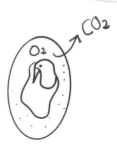

As a chick grows, it uses oxygen from the egg's air sack, and replaces it with CO_2, that is let out by tiny pores, that also let fresh air in. A chicken egg has more than 7,000 pores to allow the embryo to breathe.

在小鸡胚胎发育的过程中，小鸡会消耗鸡蛋气室中的氧气，然后产生二氧化碳。二氧化碳通过微小的气孔排出，同时也让新鲜空气进入。一个鸡蛋有超过 7 000 个气孔以便胚胎呼吸。

Are you sure that the chicken came first?

你确定先有鸡吗?

Would you be able to have a conversation about life with a predator wanting to eat you?

你能和一个想要吃掉你的天敌探讨生命吗?

What do you do when pressed for an answer you do not know? Guess?

当你被追问一个你不知道的问题时,你会怎么做? 猜猜看?

How can you relax when under stress?

如何在压力下放松?

This question about the chicken and the egg has been asked for millennia. Ask your friends and family members if they have a clear opinion on this matter. Challenge their responses, and put pressure on them to follow a chain of logic that will put it all into context. When you think that you have a clear idea yourself, summarise your findings and present these to them. Have your ideas organised, with logic in your mind, so you do not need any written notes. After you have presented your case, open the floor for questions. Enjoy the debate!

关于鸡和蛋的问题，人们已经争论了几千年了。问问你的朋友和家人，他们对这个问题是否有明确的看法。对他们的回答进行挑战，给他们压力，让他们按照一系列的逻辑，理出个来龙去脉。当你认为自己有了清晰思路时，总结一下你的结论，并向他们展示。用头脑中的逻辑梳理自己的想法，不要做任何书面笔记。说完你的论点后，放开让别人提问。辩论快乐！

学科知识
Academic Knowledge

生物学	蛋是一种具有膜结构的容器，里面有胚胎；鸡蛋因其胆固醇、脂肪和蛋白质而被狐狸当成宝贝。
化 学	蛋壳的主要成分是碳酸钙；特异蛋白质OC-17能加速蛋壳的形成，实现24小时内产出鸡蛋的需要。
物 理	蛋壳曲面越尖锐，蛋越坚硬；细胞器以及某些病毒的外壳是卵圆形的；卵圆体最坚固的部分是它狭窄的尖端。
工程学	旋转生鸡蛋（里面是液体）和熟鸡蛋（里面是固体）产生的动量和惯性；牛顿第一定律的应用。
经济学	1990年至2010年期间，家禽产量翻了一番（只数），绵羊和山羊数增加了20%。
伦理学	学习祖先的历史有助于理解他们所面临的挑战，让我们以关爱和同情的心态看待他们的缺点和错误；我们能接受圈养鸡在34天内成熟吗；在鸡饲料中添加非天然食品。
历 史	考古证据表明，大约在公元前7400年，在中国河北省磁山遗址，原鸡被驯化；亚里士多德在公元前4世纪就讨论过"先有鸡还是先有蛋"的问题；公元5年，马克罗比乌斯宣称这个问题很重要。
地 理	中国和印度尼西亚是世界上养鸡最多的国家。
数 学	产蛋需要24小时，孵化需要21天，正常条件下，鸡成熟需要100天，但是如果使用杂交技术和特殊喂养方法，成熟时间可以缩短到34天；蛋是卵圆体——一种三维椭圆体；笛卡儿和卡西尼的卵形线。
生活方式	家谱的建立。
社会学	作为一种暗喻的蛋鸡之问；一种既有的偏见认为鸡是愚蠢的，但事实上鸡会数数，并且确实显示出一定程度的自我意识，甚至会用狡猾手段操纵彼此。
心理学	蛋被视为生命的象征，因为它蕴含着初始的生命，一个新生命将从蛋中诞生；蛋体现了生命循环中再生和返老还童的理念，它的形状也象征了这样的思想——没有起点也没有终点；在梦境中，蛋代表生活的新希望，代表你想要挖掘的隐藏潜力；改变想法的能力；承认和分享令你印象深刻的事物可以增强你的信心；使人坚持和不屈的精神力量。
系统论	生命的无限循环和进化。

情感智慧
Emotional Intelligence

狐狸

狐狸自然而然地开始了谈话，以一种简单而又尖锐的方式提出了一个问题。她明确表示，除非得到满意的答案，否则不会罢休。她在接受了公鸡的第一个答案之后，又重复了这个问题，并迫使公鸡重新思考这个问题。当公鸡感到压力时，她的态度似乎有所缓和，要求公鸡放松，但随后又侮辱公鸡。当公鸡准备放弃时，狐狸又一次重复同样的问题来施加压力。当公鸡终于得出一个聪明的结论时，狐狸才感到满意，并谦恭地承认她为公鸡的逻辑和推理所折服。

公　鸡

公鸡礼貌而谦恭地回答说，别人告诉他红原鸡是现代鸡的祖先，但他没有自称是他们的祖先。当第一次被问"先有鸡还是先有蛋"时，他闪烁其词，并试图改变话题。当被追问时，公鸡说出了一个显而易见的事实，那就是他是从蛋里孵出来的，希望就此打住。当狐狸再次问这个问题时，公鸡又认为一定要有鸡才能下蛋。公鸡渐渐承受不住压力，并希望其他动物，如猫头鹰或郊狼，来回答这个问题。他要求停止提问。他知道其他动物并不认为鸡聪明，但他指出鸡确实有逻辑思考能力，也确实有同理心和自制力。公鸡最终提出了令人惊讶的新见解，彰显了他的智慧。

艺术
The Arts

美国艺术家威廉·霍尔布鲁克·比尔德有一幅著名的画作，名为《校规》，画中有一群在学校上学的动物在校长面前排成一排。你想扮成画中的哪种动物？想象一下，你会和你的（动物）同学们讨论什么？试着写一段剧本，并上台演出。

思维拓展
Systems: Making the Connections

先有鸡还是先有蛋，这是一个老生常谈的话题，但很少有人会花时间去弄清这个谜题的真相，更深入地探索生命随时间而进化的过程。对这个概念的进一步研究提供了一个很好的训练逻辑的机会，并让我们不再满足于显而易见的答案。现实是，对于如此深刻的问题，没有唾手可得的答案。我们需要时间来处理一系列基础的思路和研究结果。鸡肯定是从蛋里出来的，可是那个蛋肯定是鸡下的。所以在此之前一定有过某种生命形式。在这里，我们有机会运用一种清晰而简单的逻辑，通过反复提出相同的问题，然后得到不同的答案，每次得出的答案都在某种程度上是正确的。这说明仅仅得到一个正确的答案是不够的。需要更深入地探讨，才能充分理解生命的本质。生命是复杂的，要理解它，就需要一遍又一遍地提出同样的问题，以确保对所涉及的所有方面都有更广泛的思考。这样做是为了掌握进化的复杂性。当动物离开海洋来到陆地上时，一个进化的转折就发生了，在陆地上有必要模拟海洋的条件。人们今天吃的蛋经历了几个阶段的进化，从两栖动物的软蛋到鸟类的硬壳蛋，其中包含复杂的营养和呼吸系统。鸡蛋是一件杰出的设计作品，它具有废物的处理机制，并让二氧化碳和氧气通过细小气孔交换。因此，对于这个古老的问题，最有可能的答案似乎是：一种两栖动物下了蛋，鸡的胚胎就在这个蛋里，第一只鸡就从这里诞生。我们可以这样推断：确实是一个已有物种的蛋先出现，然后才有了鸡。

动手能力
Capacity to Implement

鸡和狐狸有可能和平相处，并对生命的哲学进行讨论吗？这只会发生在故事中，但这个问题确实让我们思考。就像古老的难题"先有鸡还是先有蛋"一样，想象一下两种不同的动物之间的交流，一种是捕食者，另一种是猎物，以及他们可能进行的对话。问问自己，这种对话的主题可能是什么，他们可能会交换什么想法和主意。我们都同意这样一个事实：哲学交流可以把人们聚在一起，了解彼此的生活和人生观，并产生同理心。在生活中，以这种方式与他人互动是一项重要技能。

故事灵感来自
This Fable Is Inspired by

罗莉·马里诺
Lori Marino

罗莉·马里诺在美国纽约大学完成了她的本科学业，1995年在新泽西州普林斯顿大学获得生物心理学博士学位。她在俄亥俄州牛津市的迈阿密大学获得了人类心理学硕士学位，之后在美国国家航空航天局（NASA）的约翰逊航天中心工作。她在纽约州立大学奥尔巴尼分校研究动物行为。之后，她在佐治亚州亚特兰大市的埃默里大学的埃默里伦理中心任教员。马里诺博士目前是金梅拉动物保护中心的执行主任，也是鲸保护区项目的创始人。她撰写了100多篇文章，内容涉及海豚和鲸的大脑进化、解剖学、智力、自我意识，以及圈养对群居动物的影响。她是《难以接受的真相：鸡会思考》一文的作者。2001年，她与人合作撰写了一份开创性的研究报告，为宽吻海豚的镜像自我认知提供了第一个确凿证据，之后她决定不再对圈养动物做任何进一步的研究。

图书在版编目（CIP）数据

冈特生态童书. 第七辑：全36册：汉英对照 /
（比）冈特·鲍利著；（哥伦）凯瑟琳娜·巴赫绘；
何家振等译. —上海：上海远东出版社, 2020
ISBN 978-7-5476-1671-0

Ⅰ.①冈… Ⅱ.①冈… ②凯… ③何… Ⅲ.①生态
环境－环境保护－儿童读物—汉英 Ⅳ.①X171.1-49

中国版本图书馆CIP数据核字（2020）第236911号

策　　划　张　蓉
责任编辑　祁东城
封面设计　魏　来李　廉

冈特生态童书

先有鸡还是先有蛋？

[比]冈特·鲍利　著
[哥伦]凯瑟琳娜·巴赫　绘

何家振　译

记得要和身边的小朋友分享环保知识哦！
八喜冰淇淋祝你成为环保小使者！